1 MONTH OF
FREE
READING

at

www.ForgottenBooks.com

By purchasing this book you are eligible for one month membership to ForgottenBooks.com, giving you unlimited access to our entire collection of over 700,000 titles via our web site and mobile apps.

To claim your free month visit:

www.forgottenbooks.com/free60728

ISBN 978-0-484-61982-0
PIBN 10060728

This book is a reproduction of an important historical work. Forgotten Books uses
state-of-the-art technology to digitally reconstruct the work, preserving the original format
whilst repairing imperfections present in the aged copy. In rare cases, an imperfection in
the original, such as a blemish or missing page, may be replicated in our edition. We do,
however, repair the vast majority of imperfections successfully; any imperfections that
remain are intentionally left to preserve the state of such historical works.

NOTES

ON THE

# of Machine Forces

BY

C. H. HECK, M.E.

PROFESSOR OF MECHANICAL ENGINEERING
IN RUTGERS COLLEGE

WITH 30 ILLUSTRATIONS

NEW YORK
D. VAN NOSTRAND COMPANY

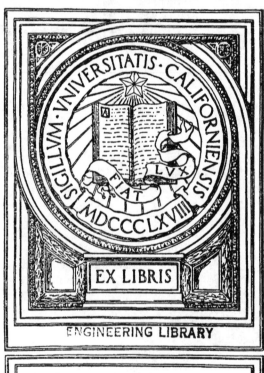

NOTES

ON THE

raphics of Machine Forces

BY

ROBERT C. H. HECK, M.E.

PROFESSOR OF MECHANICAL ENGINEERING
IN RUTGERS COLLEGE

WITH 39 ILLUSTRATIONS

NEW YORK

D. VAN NOSTRAND COMPANY

23 MURRAY AND 27 WARREN STREETS

1910

Gift
Joseph Le Conte
to
Engineering Library

THESE notes are intended to serve as text for a graphical course which has hitherto been based upon Herrmann's " Graphical Statics of Mechanisms." To the latter text there are the objections that it does not set forth the fundamental mechanical principles clearly enough for students with the usual degree of effective preparation, and that it wastes entirely too much space on detailed and repeated explanations of examples. The problems in this course are in the shape of good-sized drawings of machines—on sheets about 20 in. by 27 in.—which are reproduced from tracings as positive prints, with dark lines on a light ground. The student is thus saved the labor of mere drawing, and at once takes up the force determination. On the drawing there is room for such special notes and suggestions as may be called for; but the emphasized purpose is to have the student think for himself, with needed help and suggestion from the instructor, and not follow a ready worked example.

In section J are added some special force constructions which are useful chiefly in the problems of graphical dynamics, but which are needed to round out the presentation of graphical methods for determining impressed forces in machines.

It is thought that the title here used is more appropriate and descriptive than " graphical statics of mechanisms."

R. C. H. HECK.

NEW BRUNSWICK, N. J.,
May, 1910.

iii

M̄ 2948

CONTENTS

GRAPHICS OF MACHINE FORCES

A. General Conditions of Problems

1. In order to determine, by the simpler methods of graphical analysis, the principal forces acting in machines, we make the following assumptions:

(*a*) The weight of the machine members, or the force of gravity acting upon them, may be disregarded.

(*b*) The force of inertia need not be taken into account: this idea involves either a slow running of the machine or a practically uniform motion of the parts.

A great many problems may be solved with quite sufficient correctness under these assumptions. If the need of greater accuracy or their increase in relative magnitude requires that these forces (which are functions of mass or of mass and motion) be included, the solution becomes much more complicated. (See Section J.)

2. For present purposes, the machine is considered as made up of rigid bodies, or of members which are fixed and definite in form so far as the forces entering into the problem are concerned. These are all impressed forces, imposed through the contact of one machine part with another. The forces which act upon one piece get at each other and come into equilibrium through the medium of internal forces or stresses within the body. We pass over all questions as to the magnitude or the manner of action of these stresses, and use simply the resultant relations among the imposed forces; which relations are the same as if the forces acted upon and met in a single material particle.

B. Force Diagram Constructions

3. In every case, the forces on any member of the machine are to be considered as in equilibrium. The requirements for this condition are as follows:

If there are but two forces, they must be equal and opposite on a common line of action.

If there are three forces, their action-lines must meet in a common point, and the forces must form a closed triangle.

If there are four or more forces, they need not all pass through one point, but they must be reducible to the three-force case.

With parallel forces, we must apply separately the requirements that the algebraic sum of the forces and the similar sum of their moments about any center shall each be equal to zero.

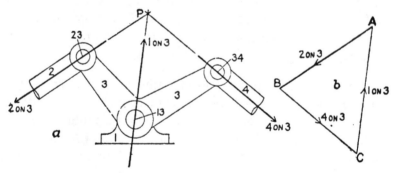

FIG. 1.—Three Forces on Bell-crank.

4. In the three-force example shown in Fig. 1, the forces on the bell-crank 3 are to be determined. Rod-pull (2 on 3) is completely known, in line of action, direction, and intensity; rod-pull (4 on 3) is known in line of action, and gives the intersection P. This point and the center of the bearing 13 determine the line of the third force, the bearing-pressure (1 on 3). Having used the drawing of the machine to find the force directions (that is, the angular positions or inclinations of the lines of action), it is often well to draw the force diagram separately, as in Fig. 1b. Here the line AB is parallel and equal to (2 on 3), BC is parallel to (4 on 3), and AC is parallel to (1 on 3). The inter-

section at C fixes the length or value of the latter two forces. Remember that for equilibrium the direction-arrows point in circular order around the triangle.

5. The two typical cases of determinate conditions under the action of four forces will now be taken up—and these will be considered as sufficiently representing all cases of the equilibrium of more than three forces on one piece. In Fig. 2, the forces on the elevator body 2 are, the load Q, the upward pull P, and the

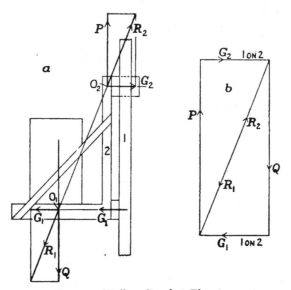

Fig. 2.—Wall or Bracket Elevator.

two guide-bar pressures G_1 and G_2;* all the action-lines and the intensity of one force Q being known. Combine the forces in pairs, Q with G_1, P with G_2. For equilibrium, the resultants R_1 and R_2 must balance, hence must have a common line of action; and the intersections O_1 and O_2 determine this line. The diagram at b shows how R_1 is found from Q and G_1, then reversed for R_2 and resolved into P and G_2.

---

* Here friction is disregarded, consequently these forces are perpendicular to the surfaces in contact.

6. In Fig. 3 the piece 4, which is made up of the wheels and axle of a locomotive, is subjected to four forces (besides its weight load, not here included); these are, the pressures of the two connecting-rods 2 and 3 upon their crank-pins, the pressure of the bearings (parts of frame 1) upon the axle, and the tangential resistance of the track T. Three lines of action are known, but for the location of the fourth force there is only one determining point. The problem is soluble when two of the forces, as (2 on 4) and (3 on 4), are completely known; whereupon their resultant R can be found, and we drop to the three-force determination, with the intersection P as the second point required to

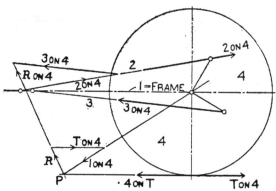

Fig. 3.—Driving Action in Locomotive.

fix the unknown force-line. Note that in this figure the force triangles are made right on the drawing of the machine; and the force R, a result in the first triangle, is moved to another place on its line of action in order to enter as a primary quantity into the second triangle.

7. The equilibrium of three parallel forces is illustrated in Fig. 4. Always, the two outer forces P and Q act in the same direction, while the equilibrant R, equal to their sum, lies somewhere between them. The only moment equations that need be considered are:

With origin on R,	$Pa = Qb$;	. . . . (1)
With origin on P,	$Ra = Qc$;	. . . . (2)
With origin on Q,	$Rb = Pc$.	. . . . (3)

All of these are expressed graphically in the triangles obtained by drawing the parallel cross-lines AC, DF, and GK, and the diagonal GF. Considering the first relation, for instance, we have from the similar triangles ABG, CBF:

$$CF:CB::AG:AB,$$

or

$$P:b::Q:a,$$

whence

$$Pa = Qb.$$

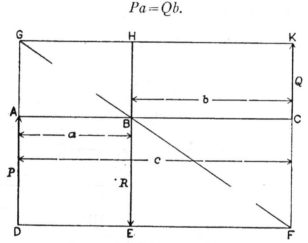

FIG. 4.—Parallel Force Relations.

For equal moments, the forces must be inversely as their own moment-arms, or directly as the opposite arms. In effect, in Fig. 4, each force is transferred to the line of the other force, P to FC, Q to AG, and there is then a direct proportion to the arms BC and AB. This idea of interchanging the forces for purposes of graphical construction is used in every case; thus for the second relation, with origin of moments on line of P, we put R at KF and Q at BH and have the forces in the ratio of the distances GK and GH. The cross-lines AB, DF, and GK need not be perpendicular to the force-lines; their parallelism is the essential thing.

8. The typical problems in parallel forces are represented in Fig. 5. In each diagram, full lines show originally known

quantities, dotted lines the results got by construction. The figures are lettered in such a manner that the alphabetical order indicates the order in drawing the lines. The four cases may be briefly set forth as follows:

Case I.—The three lines of action and an outer force P are known: draw AB and CD, to carry P over to line of Q; diagonal DEF determines Q in AF and R in CF, and line FG transfers these forces to their proper lines.

Case II.—The three lines of action and the inner force R are known: draw AB-AC and DE-DF, to carry R to line of Q;

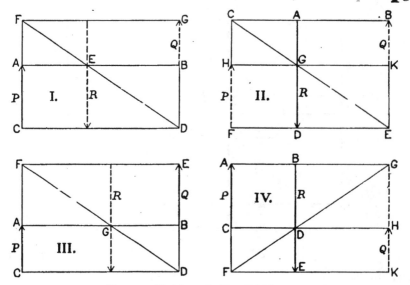

Fig. 5.—Problems in Parallel Forces.

diagonal CGE divides R into AG or Q and GD or P, and line GH-GK transfers the forces to their own lines.

Case III.—Two forces in same direction are completely known, and third (middle) force (known in intensity, as the sum of the other two) is to be located: draw AB, CD, and EF, to carry forces to opposite lines, P to DB, Q to AF; then diagonal FGD locates at G an origin about which P and Q have equal moments, which is therefore a point on line of R.

Case IV.—Two forces in opposite directions are completely known, and the third (the outer force beyond the larger of the two, and known in intensity as their difference) is to be located: draw AB, CD, EF, and diagonal FDG to meet AB produced at G; the latter point locates the line of Q, and DH and EK carry the force over to this line.

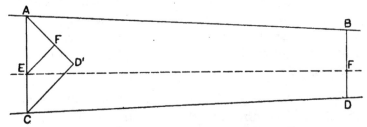

Fig. 6.—Construction for Distant Intersection.

9. When two forces are nearly parallel, so that their intersection falls far off the drawing board, it is necessary to find the direction of the third force without actually drawing it to the intersection. In Fig. 6, AB and CD are the known force-lines, and the third force is to pass through the point E. Draw a cross-line AC through E and, at a convenient distance, a parallel cross-line BD. The third force-line EF must divide BD in the same ratio as AC. From A draw AD' (at any convenient angle) equal to BD, join CD', and draw EF' parallel to CD'; then lay off point F' on BD, at F, and draw EF.

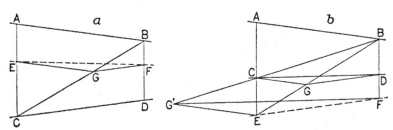

Fig. 7.—Geometrical Constructions.

10. Another scheme for dividing the second cross-line in the same ratio as the first, which involves no measurement or transfer of lengths, is shown in Fig. 7. For the first case, with the point E between the known lines, draw the diagonal CB; then EG, parallel to AB, carries the ratio (AE:EC) to (BG:GC); and drawing

GF parallel to CD makes (BF:FD) equal (BG:GC). To apply this exact scheme when E lies outside the known lines, we should have to draw EG′ parallel to AB, meeting CB produced, then draw G′F parallel to CD. It is more convenient, and tends to better accuracy, again to make the construction on the middle line, now the known line CD: draw CG parallel to AB, join GD, and draw EF parallel to GD. The essential thing is to have a line parallel to the base of a triangle, so as to divide the sides in the same ratio.

11. When three intersecting forces are not far from parallelism, the force triangle becomes very flat, as at diagram b in Fig. 8, and the intersection U of the sides representing the unknown

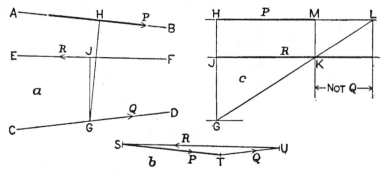

FIG. 8.—The Flat Force Triangle.

forces is rather indefinite. For accurate determination of these forces, the moment relation must supplement the force triangle. One method is to take an origin on the line of one unknown force, as at G, draw and measure the perpendicular moment-arms GH and GJ, measure the force P, and multiply and divide in the relation,

$$Rb = Pc, \qquad . \quad . \quad . \quad . \quad . \quad . \quad (4)$$

Or, as indicated at c, a regular parallel-force construction may be made for one force, as R. In either case, after a second force has been found we go back to the force triangle and use it to get the third force.

C. The Action of Friction

12. In Fig. 9 is represented a block 2 which is pressed down upon the plane 1 by a force N, normal to the surfaces in contact, and is supported by the equal and opposite upward reaction of the plane 1. If the pressure between the surfaces is uniformly distributed over the contact-area, as indicated by the diagram at the right, the equivalent concentrated force N will act at the geometrical center of this area. If the block is made to move, as in the direction of the motion-arrow, there will be developed between the surfaces a frictional resistance F, opposing the movement. This frictional force F bears to the normal pres-

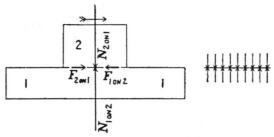

FIG. 9.—Resistance to Sliding.

sure N a ratio which is known as the coefficient of friction. As in every case, there is here the inevitable condition of an equal and opposite action and reaction: the friction force $F_{1 \text{ on } 2}$ opposes the movement of 2 on 1, while the reaction $F_{2 \text{ on } 1}$ similarly opposes the movement of 1 relative to 2.

13. In Fig. 9 only the conditions right at the contact surfaces are considered, and nothing is shown as to the detail of the external forces on the block 2. The complete representation of a simple case is given in Fig. 10. Block 2 is pulled downward by its own gravity force W, which necessarily passes through the center of gravity G; and the force P, which moves the block, acts along the line CD, parallel to the plane. To bring the block just to the eve of motion, or to maintain a uniform motion, P must equal F; and the resultant of W and P is the pressure

of 2 on 1, balanced by that of 1 on 2.   The latter force has $F_{1 \text{ on } 2}$ as one component, while the other is N, equal and opposite to W.   Since P and F are not along the same line, they form a turning couple; but their moment Pa is balanced by the equal and opposite moment Wb of the couple W and N.   Further, since the total pressure (2 on 1) or (1 on 2) is not central at the contact surface, the actual distributed force will be non-uniform, as shown by the pressure diagram at the right.

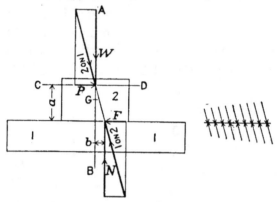

FIG. 10.—Block Sliding on Plane.

14. The arrangement outlined in Fig. 11 is frequently used to illustrate the action of sliding friction, and serves especially to give fuller meaning to the expression "angle of friction." The gravity force W is resolved into components, N normal and P parallel to the plane; and the inclination of the plane is just enough to make the active component P equal to the frictional resistance F—by "active" component is meant the one in the direction in which the body is capable of moving under its conditions of constraint of motion.   The angle α, which the plane AB makes with the horizontal base BC, or which the total pressure W makes with the normal force N, is called the angle of friction; and the coefficient of friction is the tangent of this angle, or

$$\mu^* = \frac{F}{N} = \tan \alpha. \quad . \quad . \quad . \quad . \quad . \quad (5)$$

---

* Greek *mu*, same as *m*.

15. Suppose that under the conditions of Fig. 9 or Fig. 10 a force P is applied which is less than the total friction μN; then only so much of the frictional resistance will come into play as is needed to balance P. As P increases, the frictional force be-

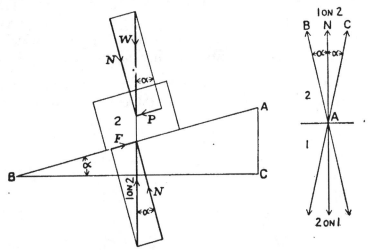

FIG. 11.—Sliding on Inclined Plane. FIG. 12.—Limits of Frictional Resistance.

tween the surfaces also increases, up to the limit μN; when this limit is reached, motion is ready to begin. If P becomes greater than F, the line of resultant pressure of 1 on 2 does not swing beyond the angle α, as shown at AB or AC in Fig. 12, but the excess of P over F is a free or unbalanced force, acting to accelerate the moving body. In a reversal of motion, as by an engine crosshead at the end of the stroke, the line of pressure swings through the angle 2α, say from AB to AC.

16. The first question to be answered in any case of sliding friction in a machine is, To which side of the normal is the pressure-line inclined? The general principle of all friction action is that friction opposes rela

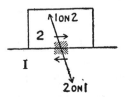

FIG. 13.—Inclination of the Pressure-line.

tive movement; consequently, the pressure acting upon any piece at the contact surface must have a component against the movement of that piece. The simple "rule of thumb" is illustrated in Fig. 13. At about the place where the force-line will cross the contact-surface, crosshatch a little block on each piece to emphasize the contact, and draw arrows indicating the respective directions of relative movement of the two pieces; then draw an inclined line which joins, or tends to join, the tails of the arrows. This line shows the direction in which the pressure-line slants away from the normal to the surfaces.

17. The machine outlined in Fig. 2 and reproduced in Fig. 14 serves as a very good example of the effect of sliding friction.

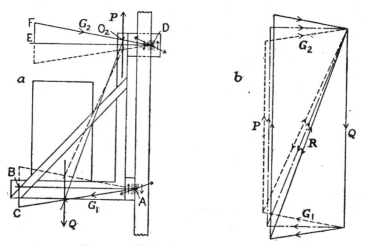

FIG. 14.—Wall Elevator, with Friction.

The first step is to sketch in at A and D the determination of Fig. 13, purposely exaggerating the angle of inclination; then the action-lines of the guiding pressures G_1 and G_2 are definitely drawn by measuring off along the normals convenient lengths AB and DE and erecting perpendiculars, $BC = \mu \times AB$ and $EF = \mu \times DE$. The intersections O_1 and O_2 now fix the resultant line, and the force diagram takes the form shown in full lines at b. With this is drawn in dotted lines the diagram for downward motion of the elevator, and in dot-and-dash lines the diagram for the ideal case of no friction, reproduced from Fig. 2. In downward movement the weight Q becomes the driving force, while P, now the hold-back or brake force, serves as resistance.

18. It is of interest to set forth a reason for the fact that, in Fig. 14, the guiding forces G_1 and G_2, and consequently the friction on the guide-bar, are greater in downward than in upward motion—that is, a reason with a more fundamental basis than the mere appearance of this result in the force-diagram determination. In Fig. 15 are given diagrams of all the vertical forces on the elevator body, the third force F being the resultant of the two frictions and equal to the difference between P and Q. The resultant turning moment of the three forces must be balanced by the couple made up of the normal components of G_1 and G_2.

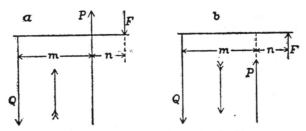

FIG. 15.—The Moment of Friction in Fig. 14.

Taking the center of moments on the line of force P, we see that in case a the moment Fn acts against Qm, while in case b these two moments act together. This example shows the advantage of using other lines of reasoning to supplement, and sometimes to check, the direct graphical determination.

D. Journal Friction

19. The turning of a journal in its bearing is nothing but a sliding of curved surfaces upon each other, but the conditions as to pressure-distribution are much less simple than with flat surfaces. In Fig. 16a the journal is supposed to fit the bearing with a snug running fit, close but not forced; then under the action of the force N the pressure would be distributed about as shown by the diagram. An actual bearing is somewhat loose

on the journal; wear in service tends toward a uniform distribution of pressure; and between the surfaces there is an elastic
film of lubricant. For these reasons the curve of distribution
will probably take the form at *b* in Fig. 16; it is more nearly
uniform over a considerable arc at the bottom, but does not cover
the entire half-circumference. The essential fact, in any case,
is that the total pressure between the surfaces is greater than
the resultant *N*—this because the oblique pressure away from
the resultant-line has only a component for or against *N*. With
the same load and the same coefficient there would be, therefore,
more frictional resistance in a bearing than under a slide-block.

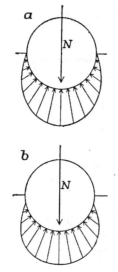

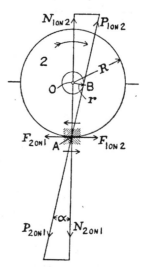

FIG. 16.—Pressure between Journal and Bearing.   FIG. 17.—Journal Friction.

20. Actually, the coefficient of friction is, for the same materials and lubricant, quite a good deal less in the bearing than on
the slide, chiefly because the lubrication is so much more effective,
as explained in paragraph 63, following. Further, the coefficient as experimentally determined includes the influence of
obliquity of surfaces, since the experiments are made with bearings
of actual form. Practically, we treat the problem as if the pressure and its resulting friction were concentrated upon a narrow

flat surface which is central on the main line of thrust. This condition is depicted in Fig. 17, where the fundamental relations are closely analogous to those in Fig. 9. It is assumed that all special details in the action of pressure and friction have exerted their influence in helping to determine the friction-coefficient ϕ (phi).

21. In Fig. 17, the pressure P makes with the normal N the friction angle α; and the coefficient is, as before, $\phi = F/N = \tan \alpha$. To oppose rotation about the journal-axis O, the frictional force F exerts the resisting moment,

$$M = F \times R = \phi N \times R. \quad \ldots \ldots \quad (6)$$

In order that the force P, which is equal to $N/\cos \alpha$, shall have this moment, it must satisfy the relation,

$$\frac{N}{\cos \alpha} r = \phi N R, \quad \ldots \ldots \quad (7)$$

and pass the center of the journal at the distance,

$$r = \phi R \cos \alpha. \quad \ldots \ldots \quad (8)$$

The same relation is easily derived geometrically; in the triangle ABO,

$$r = R \sin \alpha = R \frac{\sin \alpha}{\cos \alpha} \cos \alpha = \phi R \cos \alpha. \quad \ldots \quad (9)$$

Since the angle α is always small, we disregard the factor $\cos \alpha$, and use for the friction circle the radius,

$$r = \phi R. \quad \ldots \ldots \quad (10)$$

With the line of total pressure tangent to the friction circle, instead of passing through the center, the pressure diagram will be distorted from the symmetrical form in Fig. 16: it will be heavier on the side where the journal surface is descending, or where, to a slight degree, the journal tends to climb in the bearing.

22. Examples illustrating the use of the friction circle are given in Fig. 18, with the simple link-work fully outlined at *a* as basis. The single force-line along the connecting link 3 is to be located, and the question is, To which side of the friction circle will this line be tangent at each joint? Without friction, the line will go through the centers of the two journals. In case *a*, the rod is in a state of tension, or is pulling down on the arm 2, hence the "contact" will be on the top of the upper pin— and on the bottom of the lower pin. Noting that the angle between links 2 and 3 is becoming less or "closing," we see that 3 will have left-hand or anti-clockwise rotation with reference to 2, or that 2 will turn toward the right on 3; and the direction arrows are drawn accordingly. A line joining the tails of these arrows, as in Fig. 13, shows to which side of the friction circle the final force-line will pass. The same determination is made at the lower joint, where the relative movement is the reverse of that at the top; and the force-line is drawn tangent to the two friction circles on the sides indicated.

23. For the preliminary determination of the side to which the true line of force is deflected from the normal, it is proper to center the contact on the force-line for the case of no friction (it is understood that the force-action without friction will have been worked out before the case with friction is taken up). Remember, however, that this shows only the direction of the deflection α, and does not fix the radial line from which this angle α will finally be measured. The friction circle expresses geometrically the condition to be met by the force-line at this one journal-bearing; if the line is tangent to the circle, it will make the proper angle α with a radius drawn to the point where it crosses the circumference. To fix the line, another determinant, outside of the single bearing, is required.

24. In parts *b* and *c* of Fig. 18 are shown two out of a number of possible variations from case *a*. At *b* there is an interchange of pin and eye between the links at each joint, as compared with *a*; but the final result, in the location of the force-line, is unchanged. At *c*, however, link 4 is made driver and the direction

of motion is kept the same, thus changing the stress in rod 3 from tension to compression; and now the force-line is reversed in tangency at both joints. The following general rules govern these and other changes:

To interchange journal and bearing or pin and eye at a joint (or to be uncertain, in a problem, as to which part belongs to which piece) will not affect the force-line at this joint, provided that the contacts and motions are correctly represented for the existing arrangement.

To reverse either direction of force or direction of motion at a joint will swing the force-line to the other side of the fric-

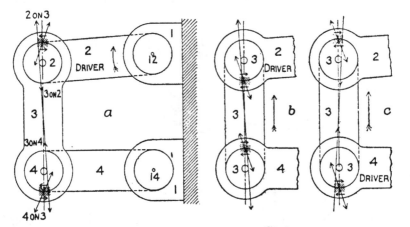

Fig. 18.—Use of the Friction Circle.

tion circle, but to reverse both together will leave it unchanged. The two changes act like minus signs in an algebraic multiplication, one effecting a reversal, two neutralizing each other.

25. A very useful check upon the detailed determination of the side of tangency to the friction circle (that is, upon the method of Fig. 17) is got by applying the general principle that friction must hinder the motion. Thus in Fig. 18a, " driver " 2 is made to turn about the center 12 by some driving force not shown on the figure, and the force (3 on 2) acts as a resistance to motion; this resistance is given a greater effect by moving it outward from 12, or increasing its moment-arm from that center.

At the lower joint, however, force (3 on 4) is a driving force, and the hindering effect of friction is shown by the decrease in its moment-arm (from 14) due to tangency on the inner side of the friction circle. This check is not always so obvious or so easy to apply as in this example, but it should constantly be kept in mind.

26. For the purpose of finding the relative motion in a turning joint, the scheme illustrated in Fig. 19 is very effective. Consider joint B: the lines AB and CB (of constant length) form two sides of a triangle, with AC as base. Whether AC will increase or decrease as the mechanism moves in the direction of the arrow,

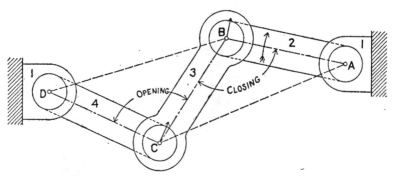

FIG. 19.—Determination of Direction of Motion.

determines whether the angle ABC is opening or closing. Point C is moving in a direction perpendicular to the arm CD; and since this direction makes an acute angle with AC, we see that the latter is decreasing. For the joint C, with DB as the variable base, the corresponding angle is obtuse, hence DB is increasing and the angle opening. If C were, for a particular position of the mechanism, moving in a direction perpendicular to AC, it would indicate that there was, for the instant, no turning at all in the joint at B.

E. The Efficiency of Machines

27. The efficiency of a machine is the ratio of the useful work delivered to the total work put in. Let P stand (as heretofore) for the driving force and Q for the useful resistance; and while P moves a distance p* let Q move a distance q*, the ratio of p to q being determined wholly by the proportions and configuration of the mechanism. Further, let the total friction of the machine be combined in a single force F, which is overcome through the distance f. Then the general relation is,

$$Pp = Qq + Ff, \quad \cdots \cdots \cdots \quad (11)$$

or, work put in equals useful work plus work wasted against friction. The efficiency is,

$$e = \frac{Qq}{Pp}. \quad \cdots \cdots \cdots \quad (12)$$

28. The graphical methods with which we are now concerned determine forces, but not distances or velocities. It is therefore desirable to be able to express the efficiency e as a ratio of forces rather than of work quantities. This end is attained with the help of the ideal case without friction, for which we have the relation,

$$Pp = Qq. \quad \cdots \cdots \cdots \quad (13)$$

Suppose that we start P as the known force, and work through to Q. The value without friction, which we call Q_0, will be greater than the actual value Q, and the efficiency will be,

$$e = \frac{Qq}{Pp} = \frac{Qq}{Q_0 q} = \frac{Q}{Q_0}. \quad \cdots \cdots \quad (14)$$

---

* These are effective distances, measured in the directions of the forces. If the force is oblique to the path of the point of application, we may use the component of motion along the force-line for p or q, or we may use the total motion and with it the force-component along the path as P or Q.

If, on the other hand, we have Q known and work back to P, the actual P (with friction) will be greater than the ideal P_0, and the efficiency will be,

$$e = \frac{Qq}{Pp} = \frac{P_0 p}{Pp} = \frac{P_0}{P}. \qquad \cdots \cdots \quad (15)$$

29. Some machines, notably hoisting machines (in which gravity acts as the load-force Q), can run backward with the same set of forces as in forward running, but with Q now acting as driving force and P as resistance. To distinguish the forces acting under this condition, we bracket them thus, (P), (Q): the work equation takes the form,

$$(Q)q = (P)p + (F)f, \qquad \cdots \cdots \quad (16)$$

and the expressions corresponding to equations (14) and (15) are,

$$(e) = \frac{Q_0}{(Q)}, \qquad (e) = \frac{(P)}{P_0}. \qquad \cdots \cdots \quad (17)$$

The forward efficiency e is always less than unity, lying somewhere between zero and one; the backward efficiency (e) may pass through zero and become negative, and on this negative side may become greater than one. This peculiar state of affairs can best be understood with the help of an example.

30. The mechanism in Fig. 20, outlined for each case at a, has for its moving parts the wedge 2 and the vertical slide-block 3, with the load Q. In forward motion, the wedge is pushed toward the right and the load lifted; in backward motion, the wedge is withdrawn and the load descends. The two cases are, I, wedge steep or blunt, backward efficiency positive; II, wedge flat or sharp, backward efficiency negative. On diagrams a the various force-directions are indicated, but the lines are drawn so as to keep clear of each other, and no attempt is made to get the proper intersections. As in Fig. 14, dot-and-dash lines are used for forces without friction,* full lines for forward-motion

---

* Without friction, the forces are the same in either forward or backward motion.

forces, dotted lines for the case of backward motion. Diagrams *b*, with all the forces marked, should require little explanation: a force triangle with Q as the known side is first drawn for piece 3; then force (2 on 3), reversed to (3 on 2), is the base for the construction of the triangle for piece 2.

31. The essential difference between the two cases in Fig. 20 lies in the opposite directions of the force (P). In case I it

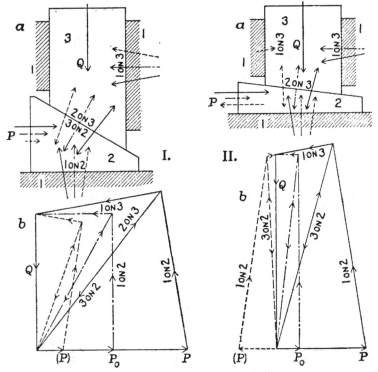

FIG. 20.—Reversal of Backward Efficiency.

points in the same direction as P_0 or P; and Q, driving the machine backward, does a " useful " work in overcoming the resistance (P). In case II the friction effect is so great, relative to the forces along the wedge, that Q alone cannot produce backward motion. Force (P), instead of holding the wedge in place and acting as a brake-force, must turn around and help Q. The full effect exerted by Q, if expressed as a horizontal force on the wedge 2, is equal to P_0. If the friction in the machine is just enough to balance Q or to neutralize P_0, the backward efficiency (e) is zero. If, as here, the friction exceeds this amount, or (P) has to help Q, this efficiency becomes negative. In a word, the work Qq is supplied at what is now the input end of the machine; but to make it move at all the further work (P)p must be supplied at the output end. In the sense of a normal output, this work (P)p is negative, hence the minus sign for (e).

32. A machine which has so much friction that it will not run backward under the action of its load is said to be self-locking. With this property goes a low efficiency in forward running. Consider the case where the friction-work (F)f is just equal to P_0p or Qq, and assume that in forward running the lost work Ff is the same as (F)f. Then from equation (11) we have,

$$Pp = Qq + Ff = 2Qq, \quad \cdots \quad (18)$$

and the value of e is one-half or 50 per cent. If (F)f is greater than Qq, e will fall below this limit. The difference between (F)f and Ff will be small, with a general probability that Ff will be the larger quantity, because the forces in the machine are likely to be a little greater in forward than in backward running. The self-locking machine has the advantages of a high velocity-ratio of P to Q—for it is because P_0 is relatively small that friction can overbalance it—and is usually a simple device for lifting a big load with a small driving force; also, it makes a safe hoist, because no brake is needed to hold up the load. The fact that at least half, and probably more, of the work put in will be expended in overcoming friction is, however, a decided drawback. The most common examples of this type of machine are the screw and the worm and worm-wheel, both derivatives from the wedge mechanism in Fig. 20.

F. Resistance to Rolling

33. When a heavy or heavily loaded roller rests on a plane surface, there is always some elastic compression of both roller and plane at the contact. This contact does not really exist along a line, but is spread over a narrow surface; the character of the pressure-distribution is indicated in Fig. 21, but with the width of contact tremendously exaggerated. If the roller advances, there is a continual compression of the surfaces on the side of advance and a release or expansion on the opposite side; the net result is a resistance to rolling, which can be most simply represented by shifting the resulting supporting force N

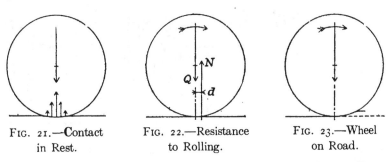

FIG. 21.—Contact in Rest.   FIG. 22.—Resistance to Rolling.   FIG. 23.—Wheel on Road.

through a small distance d toward the side of advance, as in Fig. 22. This force now passes the center of the roller at the distance d, and exerts the moment Nd or Qd, against the turning. If instead of a roller on a firm, smooth, clean track we have a wheel on a road covered with dust or mud, sand or loose stone, this effect is very much greater, as indicated by Fig. 23.

34. A long-accepted approximation for the value of d is that it may be taken as a constant at 0.02 inch, being independent of the diameter of the roller, and also of the material, provided that the loading is done with due regard to the strength of the latter. The low resistance obtained with small diameters in ball and roller bearings indicates much smaller values of d for these arrangements. We shall not attempt any general discussion of this question here; but for the problems in the present

course shall assume the constant value $d=0.02$ inch. Note that if the machine is drawn to a reduced scale, the distance d must be divided accordingly.

35. The typical conditions under which rolling resistance will enter into a graphical determination are represented in Figs. 24 to 26. In Fig. 24 the driving force P passes through the center of the roller and is parallel to the track; then at the track surface there is developed an equal and opposite holding friction T, which is a resistance to slipping at the contact; and the moment PR is equal to Qd. The diagonal OC is the common line of the resultants of Q and P and of N and T, and serves to determine the value of P in force diagram b.

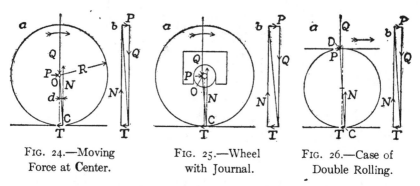

FIG. 24.—Moving   FIG. 25.—Wheel   FIG. 26.—Case of
Force at Center.   with Journal.   Double Rolling.

Fig. 25 shows the actual wheel, with the load Q impressed through a bearing and journal. The force P must now overcome both rolling resistance and journal friction, and the resultant line OC has the added inclination due to the radius of the friction circle, measured horizontally toward the side away from d.

In Fig. 26 is represented the case of double rolling, the roller lying between two flat surfaces; this is typical of ball and roller bearings. The displacement d is now measured off at both contacts, in opposite directions, but the resultant line has the same inclination, and the force P the same value relative to Q, as in Fig. 24. This force is the same, with double resistance, because its moment-arm is twice what it was in Fig. 24; the equation of moments is now $P\times 2R=Q\times 2d$. From another point of view, note that in one revolution of the roller P moves $2\pi R$ in Fig. 24 and $4\pi R$ in Fig. 26, doing twice as much work in the second case, and thus overcoming the two rolling resistances, each over the angular distance 2π.

G. Toothed-gear, Chain, and Rope Resistances

36. The fundamental conditions as to the transmission of pressure by gear teeth are set forth in Fig. 27. With involute

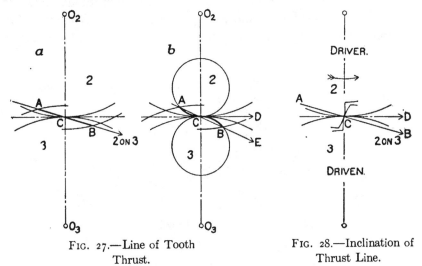

FIG. 27.—Line of Tooth   FIG. 28.—Inclination of
Thrust.   Thrust Line.

profiles, case a, the locus of the point of contact is a straight line AB, at an angle of about 75 deg. * with the line of centers, and this is also the line of tooth thrust. With cycloidal profiles, case b, the corresponding locus is a double curve ACB, made up of arcs of the two describing circles; and the line of thrust is continually oscillating from CD at 90 deg. to CE at perhaps 60 deg. with the line of centers. For graphical purposes we assume either that gears have involute teeth, or that the action of cycloidal teeth may be well enough represented by an average line of thrust, constant at 75 deg.

---

* In some systems the angle is not exactly 75 deg., but here we shall use this angle.

37. Fig. 28 illustrates a simple rule for fixing the direction in which the line of thrust is deflected from the tangent to the pitch circles. If from the pitch point C we run out, in the direction of motion, along the tangent CD, the thrust-line AB— or, to be absolutely exact, the part CB—will be swung away from the driver and toward the driven gear. A few trials of possible cases, with sketched-in profiles of the sides of teeth in action, will show that this always holds.

38. The teeth of a pair of gears slide together as they approach the pitch-point, slide apart as they recede from it. The friction action which accompanies this sliding is illustrated in

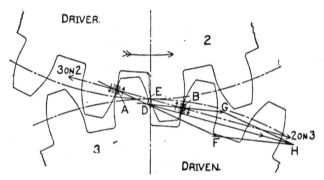

FIG. 29.—Friction on Gear Teeth.

Fig. 29, which is supposed to represent general, average conditions. The dot-and-dash line AB, passing through the pitch-point, is the line of thrust for the case of no friction. At the contacts A and B are made the regular determinations for the deflection due to sliding friction, resulting in the pressure-lines AF and BG. These lines meet at D; and the pressures being assumed equal and laid off in DF and DG, have a resultant DH which is parallel to AB.

39. The distance DE, through which the line DH is shifted from AB, is found as follows:

Angle DAB or DBA is the friction-angle α; distance AB is essentially the same as the circular pitch of the teeth, which we shall call t; and letting $\mu = \tan\alpha$ be the coefficient of friction as heretofore, we have,

$$\mathrm{DE} = \tfrac{1}{2}t \times \tan\alpha = \tfrac{1}{2}\mu t = s. \qquad . \qquad . \qquad (19)$$

The general principle set forth in paragraph 25, that friction must hinder motion, decides that this shift s will always be away from the driver and toward the driven wheel; thus giving, in Fig. 29, a greater lever-arm to the resistance (3 on 2), and decreasing the lever-arm of the driving force (2 on 3).

40. If a chain carrying a load runs over a sheave or pulley, as in Fig. 30, there is developed a certain amount of frictional

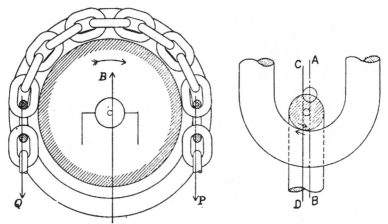

FIG. 30.—Friction in Hoisting Chain.

FIG. 31.—Action of Friction between Links.

resistance due to the turning of the links upon each other at the places where the chain runs on and off the pulley. At the " on " side, the tension Q is a load, and the friction causes it to be moved outward, farther from the center of the pulley; at the " off " side, the tension P is the driving force, hence it is shifted inward and its moment-arm decreased; and to agree with these effects, friction at the journal causes the bearing-pressure B to be moved away from Q and toward P. Fig. 31 shows that the turning of one link in another is that of a pin

within a very loose eye; and that there is a tendency for the pin
to hang or climb on the side of the bearing, thus giving to the
force-line a deflection greater than the radius of the friction
circle for the pin. The amount of the sidewise motion is determined by the requirement of making the force-line come
tangent to two friction circles, one for pin and one for eye. This
effect can be, and is, most simply taken into account by using
an abnormally large coefficient of friction, with the diameter
of the round bar, or " pin," as that of the equivalent close turning
joint.

41. The condition just described (represented in Fig. 30)
belongs to the case where the pulley merely changes the direction of the chain; except for friction, the tensions P and Q are
the same. When, however, the chain exerts a turning effort
on the pulley (or vise versa), there is an action analogous to the
sliding of gear teeth, and a corresponding friction effect. With
a round-bar chain, pockets are formed in the wheel-rim, into
which the links settle; with flat-link transmission chains, toothed
sprocket wheels are used. If chain and wheel fit properly, they
will run smoothly, and the friction on pocket-edges or teeth
can be added to that of the chain joints, merely increasing the
displacement of the force-lines. If through wear in its joints
the chain stretches, the tooth action will become very irregular,
and only the average effect can be, perhaps, roughly calculated
or represented.

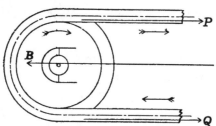

Fig. 32.—Rope Resistance.

42. When a rope runs around a pulley, as depicted in Fig.
32, there is at the " on " and " off " points a bending or straighten-
ing which can take place only by the sliding of the fibres upon

each other, and which is resisted by the friction of these fibres. The effect is to shift the " on " force Q outward and the " off " force P inward, just as in Fig. 30. The displacement, for hemp ropes, is given by the empirical formula,

$$s = 0.23d^2, \qquad \cdots \cdots \cdots \quad (20)$$

d being the diameter of the rope in inches. Note that this s must always be worked out with the full-size value of d, not with d as scaled directly from a reduced drawing, and then brought down to scale. The value of s will vary with the kind of rope, with age and condition, and perhaps with the tension; but for the problems in this course the formula just given will be considered as of universal application.

H. Belt Transmission

43. The manner in which power is transmitted by a belt is represented in Fig. 33. In a state of rest the belt is simply stretched around the pulleys, and there is the same tension T_0 on both " sides." When pulley 2 drives pulley 3, the tension in the tight side (which approaches 2) rises to T_1, while that in the slack side (running toward the driven pulley) falls to T_2. The effective force transmitted is the difference of tensions,

$$P = (T_1 - T_2). \qquad \cdots \cdots \cdots \quad (21)$$

With moderate loading, or so long as P remains well within the capacity of the belt, the sum of the tensions remains nearly constant, or,

$$(T_1 + T_2) = 2T_0; \qquad \cdots \cdots \cdots \quad (22)$$

but if the belt is overloaded and made to slip at a considerable rate, this sum increases. The effective radii for these forces should be measured to the middle of the belt thickness: with crowned pulleys, it is usually accurate enough to take the radius of the pulley at the middle of its width. Strictly, there is a stiffness action in a belt, like that shown for a rope in Fig. 32; but the effect is relatively insignificant, unless with thick belts on small pulleys, and we shall disregard it.

44. In Fig. 33 (and in all our present problems) it is assumed that the belt is stretched to a straight line, or that there is no sag due to weight. The two belt lines are prolonged to their intersection at A, from this point the tensions T_1 and T_2 are laid off, and the resultant is found. This resultant is the pressure on the bearings, due to belt pull; also, it exerts on each pulley the same turning moment as does the effective force P at the circumference. In any problem in belt gearing we wish to locate and determine this resultant; to do so, we need to know the ratio of T_1 to T_2.

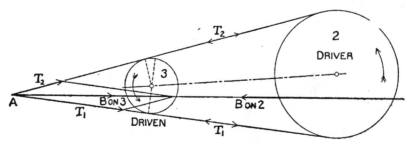

Fig. 33.—Action of Belt Drive.

45. The belt is kept from slipping around on the pulley by friction—but this is a holding friction which prevents, not merely opposes, relative movement. Referring to Fig. 12 and paragraph 15, we may compare belt friction to the holding friction, less than μN, which balances a force that is not large enough to overcome the total friction and produce motion. The conditions as to pressure and friction between belt and pulley are illustrated in Fig. 34. Diagram a reproduces the driven pulley from Fig. 33. Over the arc from the on point A to the off point B the tension increases at a continuous rate from T_2 to T_1; and at any position located by the angle α its value may be represented by the general (variable) symbol T. The unit pressure n between belt and pulley will vary with T, as will the friction μn; and the total friction developed will be equal to $T_1 - T_2$.

46. Consider now conditions over an infinitesimal arc $Rd\alpha$, as represented in Fig. 34b. The normal pressure dN on this length (or area) is, from the force diagram drawn,

$$dN = [T + (T + dT)] \sin \tfrac{1}{2}d\alpha$$
$$= 2T \sin \tfrac{1}{2}d\alpha = Td\alpha, \quad \ldots \ldots \quad (23)$$

the simplifications of the expression being exact at the limiting, infinitesimal condition. The friction μdN is equal to dT, whence we have,

$$dT = \mu T d\alpha. \quad \ldots \ldots \quad (24)$$

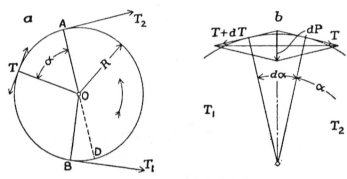

FIG. 34.—Action of Belt Friction.

Integrating from zero to α, or from OA to OB on Fig. 34a, we have,

$$\int_{T_2}^{T_1} \frac{dT}{T} = \log_e \frac{T_1}{T_2} = \mu\alpha;$$

or,

$$\frac{T_1}{T_2} = e^{\mu\alpha} = r. \quad \ldots \ldots \ldots \quad (25)$$

Here e is the base of the Naperian logarithms, and α is expressed in radian units.

47. The relation given in equation (25) is put into graphical form by means of the logarithmic spiral drawn in Fig. 35. The

radius OA is taken as unity, since for $\alpha=0$ the ratio $r=1$; and the origin of the curve is at OA.   For the coefficient of belt friction, $\mu=0.28$ is used as a good average value.   Any radius vector OB shows the value of r for the contact-angle AOB. With a pair of unequal pulleys the arc of contact on the smaller is what determines the ratio of T_1 to T_2; with an open belt (under the assumption of no sag) this arc will never exceed a

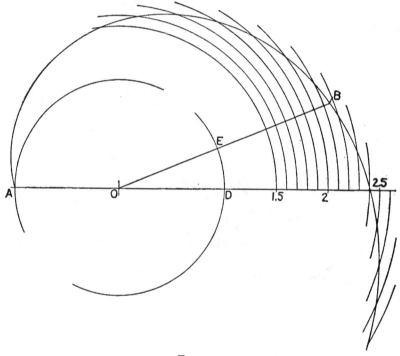

FIG. 35.

half-circle, but with a crossed belt it will be greater.   In locating a radius OB on Fig. 35, it is well to measure off the acute-angle supplement DOB (see also Fig. 34a) rather than the obtuse angle AOB.   Generally, the ratio thus found will be used to fix the line of the resultant belt-pull, and after that force has been determined it will be resolved back into components in order to get the individual tensions.   To locate the resultant-line, lay off any convenient length on the line of T_2 and r times this length on the line of T_1, and complete the parallelogram, as in Fig. 33.

48. What has been given is only the beginning of a complete theory of belt action, but it is enough for present purposes. In all problems involving belt gearing, we will assume that the relation of load to tensions corresponds to an existing friction-coefficient of the value 0.28, and will use Fig. 35 to find the ratio of tensions, after the manner just indicated. The value of r can be read off accurately enough (to two places of decimals) with the help of the circular scale-lines on the diagram. At 180 deg. the value of r is 2.41.

I. General Procedure

49. In solving a problem which calls for the determination of the forces in a machine, proceed as follows:

(a) Begin with the piece on which the known force acts, whether driving force P or load force Q. Determine the forces on this piece, then go to the next, and so on till the last piece is reached.

(b) In considering any piece, note the other machine-members which touch it in sliding or turning joint or other contact, and remember that each of these impresses a force upon the piece under consideration.

(c) Go through the whole force construction without friction first. On the diagrams mark clearly the directions of all forces, showing for each piece, in the individual triangle or portion of the diagram, the forces which act upon the piece, not its actions outward.

(d) Having thus found the general directions of all forces, mark the " contacts " after the manner of Fig. 13 or Fig. 17, determine and mark the relative movements (by inspection or, if necessary, by the method of Fig. 19), and show the inclinations of the force-lines with friction.

(e) In getting friction-circle diameters by equation (10) and in calculating the displacement of gear-tooth pressure by equation (19), the scale of representation may be disregarded, and journal radius or tooth pitch be used as measured directly from the drawing. But for rolling resistance and rope stiffness, note the remarks in paragraphs 34 and 42.

(*f*) Belt friction is not classed with the harmful or wasteful resistances, and in the no-friction case is supposed to have its full effect. The same resultant-line is used both without and with friction in the machine.

(*g*) Work should be accurate and neat, the lines being drawn with a hard pencil, sharpened to a fine chisel-edge. See that the T-square is straight. Measure as accurately as possible. A triangular scale with graduations from 10 to 60 per inch will be found most convenient.

(*h*) Use judgment in placing force diagrams. Sometimes it is better to draw them right on the figure of the machine, sometimes separate diagrams are preferable. Use different kinds of lines for the forces, as in Figs. 14 and 20, namely, dot-and-dash for no friction, full lines for forward motion, dotted lines for reversed motion.

J. Special Force Constructions

50. In contrast with the forces impressed upon a machine piece by its neighbors, such forces as weight and inertia, dependent upon the mass of the body itself, may be called self-developed. The action of gravity is readily known, but is usually of minor magnitude, except in very heavy and slow-moving machinery; the determination of inertia forces, often of large magnitude, is more difficult, and belongs to the subject of dynamics, which logically comes after the present course. We shall now briefly consider several methods which are especially useful for problems involving forces of this class, but which properly come here as extensions of the methods given in section B. It must be clearly kept in mind that so far as graphical or mathematical relations are concerned, the kind of force which acts is immaterial; the constructions now to be presented are of universal applicability, but are most likely to be called for in problems into which these non-impressed forces enter.

51.   In Fig. 36, AB or 2 and BC or 3 are two links or members of a mechanism, and upon them act the known transverse forces F_2 and F_3 respectively.   We wish to determine the other forces on these links, for equilibrium, namely, the pin-pressures or impressed forces at A, B, and C.

For link 2 take an origin of moments at A; then the moment M_2 of the force F_2 about A must be balanced by the moment of the unknown force (3 on 2).   Similarly, with an origin at C for link 3, the moment M_3 of F_3 must be balanced by that of (2 on 3).   Since (3 on 2) and (2 on 3) are equal forces, it is required that their line of action pass the centers A and C at

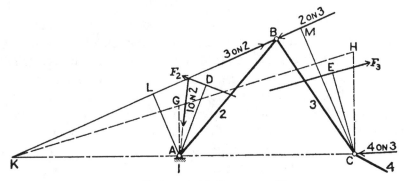

FIG. 36.—Linkage with Transverse Forces.

distances which are respectively proportional to the moments M_2 and M_3.   To find this line, proceed as follows:

By measurement and computation, get numerical values for these moments, and to a convenient scale lay off $AG = M_2$ and $CH = M_3$, making these two lines parallel to each other and somewhere near perpendicular to AC.   Join CA and HG, and produce them to meet at K; then KB is the force-line sought, since, with AL and CM drawn perpendicular to KB,

$$AL:CM = AK:CK = AG:CH = M_2:M_3. \quad . \quad . \quad (26)$$

Having found this line of force at joint B, we get determinate conditions for both pieces, 2 and 3.

52. In using Fig. 36 there must be a clear reason for laying off AG and CH either on the same side or on opposite sides of AC. Since (3 on 2) and (2 on 3) are opposing forces, their moments will be opposite if the arms AL and CM are on the same side of AC, but alike if these lever-arms are opposite: consequently, if the original forces have opposing moments as in Fig. 36, the lines AG and CH are to be drawn in the same direction; but if the moments M_2 and M_3 tend to produce rotation in the same direction, their representing lines must be opposite. The distant-intersection constructions of Figs. 6 and 7 will often be found useful in this connection.

53. Referring to Fig. 1 and paragraph 4, we see that in order to have a determinate condition for a body under the action of three forces we must know one force completely, the line of another, and at least a point (not the intersection) through which the third force must pass. In Fig. 36 we have, for each link, one force fully known, but only a point on the line of each of the other two; and in the example there presented two such indeterminates are combined to produce a definite condition. It is now in order to consider a little more fully this combination of one known force and two points of application.

54. In Fig. 37, the link or rod AB is acted upon by the known force F, while the other two forces are to act through, or be applied at, the pin-centers A and B. The lines of these two forces must meet on the line of F, but the intersection may be anywhere on that line; there is, consequently, an infinite number of possible pairs of such forces. In the figure, two points are chosen, at D and at E, and for each the force F is measured off and resolved, to get its components or equilibrants at A and B. These forces, P_A and P_B, are then carried out to A and B, and laid off from these respective points, on the lines of action. Further, the forces are at A and B resolved into two components, one parallel to F, the other along the line of centers AB; and the very important relation appears that the parallel components AM and BN are constant, while those along the line AB balance each other.

55. By drawing GK parallel to AB in the force triangle DGH, and noting the similarity of triangles DGK, DAC, and of HGK, DBC, we have,

$$DK : DC : : GK : AC,$$

and

$$HK : DC : : GK : BC;$$

from which we can get,

$$DK : HK : : BC : AC. \quad . \quad . \quad . \quad . \quad (27)$$

Since the lengths AC and BC are fixed, it follows that the constant force F is divided at K in a fixed ratio; therefore the parts,

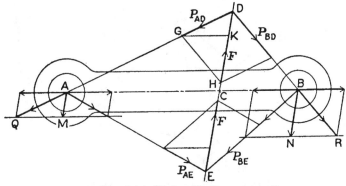

Fig. 37.—Two-joint Link with Transverse Force.

DK equal to AM and HK equal to BN, are constant. Further, since AM and BN are in the inverse ratio of their distances AC and BC, they are, as is apparent on the mere inspection of the diagram, nothing but the parallel components of F at A and B, or the equilibrants of those components.

56. What has just been shown by geometrical analysis can, perhaps, be more simply set forth as follows:

The forces applied at A and B must balance F and also balance each other, exerting one effect parallel to F, the other along the line AB. Each one must be, therefore, compounded of the equilibrium component of F, either AM or BN, and a component along the line AB. If we draw a line through M parallel to AB, all vectors from A representing possible values of the pressure P_A will have their ends on this line MQ; and NR is a similar locus for P_B.

Having then the force F on the link AB, we divide F into parallel components at A and B; and lines drawn through the ends of these forces and parallel to the center-line express, in graphical form, the requirement which the applied forces P_A and P_B must meet in order to produce equilibrium. The force-action on the body is still indeterminate, but it is now in shape to respond at once to the imposition of a determining requirement from without.

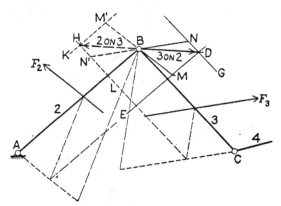

FIG. 38.—Method of Fig. 37 Applied to Problem in Fig. 36.

57. Fig. 38 shows how the locus device just described can be used for finding the forces at the joint A in Fig. 36. Consider force (3 on 2): it is a partial equilibrant of F_2, and if we measure off the component BM against the direction of F_2, force (3 on 2) must be a vector from center B to some point (not yet known) on the line ME. Again, force (3 on 2) is a direct component of F_3, wherefore BN is laid off in the direction of F_3, and (3 on 2) must be a vector from B to some point on the line NG. The intersection at D gives a vector which satisfies both requirements, and thus fixes the force (3 on 2). The dotted loci HK and HL show the same determination for the

force (2 on 3). This scheme involves less graphical work than that of Fig. 36, but confusion as to the sides of the center-lines on which to put the loci must be avoided. It is best to use this construction, and to check it by the relation, based on paragraph 52, that if F_2 and F_3 have opposite moments about A and B, the line of (3 on 2) will be outside the angle ABC; but if these moments are alike, the force-line will lie within the angle ABC.

58. A final example, in Fig. 39, will sufficiently illustrate the use of the method developed in Fig. 37. In the engine mechanism there shown, the pressure of the wrist-pin 2 on the

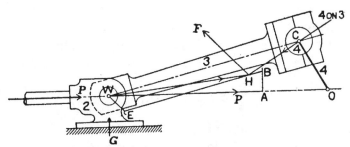

FIG. 39.—Forces in Engine Mechanism.

connecting-rod 3 will be the resultant of the known driving force P and the unknown guide-bar pressure G. The weight-and-inertia force F on the rod is known. The locus EB for the rod expresses the internal requirement for equilibrium. Laying off P as WA and drawing AB parallel to G, the external requirement is expressed by saying that (2 on 3) must be a vector from W to some point on the line AB. The intersection at B gives the force that satisfies both requirements.

59. The scheme of Fig. 37 would apply equally well if the force F were impressed by another machine piece; and a three-joint piece with a force such as weight or inertia can be reduced to the case of Fig. 37 by combining the latter force with one of the impressed forces, which at the least must be known if conditions are to be determinate.

60. The methods of Figs. 36 and 37 are really effective only
for the case of no friction.   When force-lines must be tangent
to friction circles instead of passing through bearing-centers,
exact geometrical relations become too complex to be useful.
A serviceable scheme, and one quite accurate enough, is as
follows:

Without friction, find the bearing-pressures in the several
joints of the mechanism.   Using the methods of kinematics as
necessary, find the relative velocities at the rubbing surfaces.
Get the several work-rates of friction, each as the product of
coefficient, pressure, and velocity, sum up for the whole machine,
and find the equivalent force at the point of application of P or
Q, to be added to P_0 or subtracted from Q_0.   This plan need
only be suggested here; its fuller development belongs to problems which involve much of the dynamics of machinery, such
as, for instance, the graphical analysis of engine governors.

K. Friction and Lubrication

61. A full discussion of this subject is inappropriate here,
but a brief statement of facts and principles is not out of place.
There have been two formulations of " law " from experiments.
According to the older " theory," based chiefly upon Morin's
results which were published about 1835, the coefficient of friction is, for given materials and lubricant, constant over a wide
range of conditions;   the coefficient and the resistance which it
measures were both said to be independent of area of contact
and of velocity.   There is the proviso that the area must not be
so small as to make the pressure excessive; and in the matter
of velocity, the resistance is greater when just starting and with
very slow motion than at a fair speed.   It is now known that
this hypothesis is good for dry, smooth surfaces under moderate
pressure and for the case of restricted lubrication under heavy
pressure; but even then the ratio of friction to pressure is a
more variable quantity than was supposed.

62. The later laws, formulated about 1885, are derived from
experiments with complete lubrication—by which is meant, with

a film of oil between the surfaces and completely separating them. The resistance now to be overcome is fluid friction, and the manner of its action is as follows:

In a given bearing the resistance increases but little with the load; or, for a given load, the resistance is almost proportional to the areas in contact. This makes the coefficient of friction vary inversely as the intensity of pressure. In regard to the establishment of this condition, everything depends upon the maintenance of the oil-film; and this, in turn, depends partly upon the strength, or body, or toughness of the oil, and partly upon the continuity of supply. As before, resistance is high at starting and decreases with speed; but presently it begins to increase with velocity, probably because the relative rate of supply is less, and the film is not so well maintained. While the oil must have enough body to carry the load without being squeezed out of the bearing, it should not have excessive stickiness or viscosity. Increase in fluidity is the reason why bearings usually run easier after they have become warmed up.

63. In actual machines, the conditions of working will lie somewhere between the extremes of friction in a constant ratio to pressure and friction independent of pressure. The frictional losses increase with the forces transmitted, but at a slower rate than these forces. If the machine is well loaded, the friction is much less with full than with restricted lubrication— that is, with flooded bearings as against a scanty supply of oil. This question of supply explains the observed fact that, with the same kind of surfaces and the same lubricant, friction is greater on a flat slide than in a bearing, as remarked in paragraph 20. In the former case, the oil is continually being squeezed out and pushed away; in the latter it is carried around the journal and fed in again on the side of approach.

64. With a moderate supply of oil, the coefficient of friction is likely to range from 0.06 to 0.10 in bearings, and two or three points higher in slides, or from 0.08 to 0.13. With full flooding and with a good accommodation of lubricant to pressure, it may be as low as 0.01 to 0.02 in well-made bearings. It is to be

noted that geometrical perfection of shape, with consequent uniformity in the thickness of the oil-film, is highly important to easy running. In the practice problems of this course there is a natural tendency to use high values of the coefficient of friction.

STANDARD TEXT BOOKS

PUBLISHED BY

D. VAN NOSTRAND COMPANY,

NEW YORK.

ABBOTT, A. V. The Electrical Transmission of Energy. A Manual for the Design of Electrical Circuits. **Fifth Edition, entirely rewritten and enlarged.** Fully illustrated. 8vo, cloth................net, $5.00

ASHE, S. W., and KEILEY, J. D. Electric Railways, Theoretically and Practically Treated. **Vol. I, Rolling Stock. Second Edition, Revised.** 12mo, cloth. 290 pp., 172 illustrations............net, $2.50

ASHE, S. W. Vol. II. Engineering Preliminaries and Direct Current Sub-Stations. 12mo, cloth. Illustrated.....................net, $2.50

ATKINSON, A. A., Prof. Ohio University. Electrical and Magnetic Calculations, for the use of Electrical Engineers and Artisans, Teachers, Students, and all others interested in the Theory and Application of Electricity and Magnetism. **Third edition, revised.** 12mo, cloth. Illustrated...net, $1.50

—— **PHILIP. The Elements of Electric Lighting,** including Electric Generation, Measurement, Storage, and Distribution. **Tenth edition.** Illustrated. 12mo, cloth.......................................$1.50

—— **The Elements of Dynamic Electricity and Magnetism. Fourth edition.** 120 illustrations. 12mo, cloth.................$2.00

—— **Power Transmitted by Electricity, and its Application by the** Electric Motor, including Electric Railway Construction. **Fourth edition, fully revised, new matter added.** Illustrated. 12mo, cloth....$2.00

AUCHINCLOSS, W. S. Link and Valve Motions Simplified. Illustrated with 29 wood-cuts, 20 lithographic plates, together with a Travel Scale, and numerous useful tables. **Fifteenth edition, revised.** 8vo, cloth...$1.50

BARNARD, J. H. The Naval Militiaman's Guide. Full leather, pocket size...$1.25

BARRUS, G. H. Engine Tests: Embracing the Results of over one hundred feed-water tests and other investigations of various kinds of steam-engines, conducted by the author. With numerous figures, tables, and diagrams. 8vo, cloth. Illustrated..........................$4.00

BARWISE, S. M. The Purification of Sewage. Being a brief accunt of the Scientific Principles of Sewage Purification and their Practical Application. 12mo, cloth. Illustrated. **New edition**............net, $3.50

BEAUMONT, ROBERT. Color in Woven Design. With 32 colored plates and numerous original illustrations. Thick 12mo, cloth.......$7.50

BEDELL, F. Direct and Alternating Current Testing. Assisted by Clarence A. Pierce. Illustrated. 8vo, cloth, 250 pp.........net, $2.00

BEGTRUP, J., M.E. The Slide Valve and its Functions. With Special Reference to Modern Practice in the United States. With numerous diagrams and figures. 8vo, cloth................................$2.00

BERNTHSEN, A. A Text-Book of Organic Chemistry. Translated by George McGowan, Ph.D. **Fifth English Edition.** Revised and extended by the author and translator. Illustrated. 12mo, cloth...$2.50

BIGGS, C. H. W. First Principles of Electricity and Magnetism. A book for beginners in practical work, containing a good deal of useful information not usually to be found in similar books. With numerous tables and 343 diagrams and figures. 12mo, cloth, illustrated.......$2.00

BLYTH, A. W. Foods: Their Composition and Analysis. A Manual for the use of Analytical Chemists, with an Introductory Essay on the History of Adulterations. With numerous tables and illustrations. **Fifth edition, thoroughly revised, enlarged and rewritten.** 8vo, cloth..$7.50

BODMER, G. R. Hydraulic Motors and Turbines. For the use of Engineers, Manufacturers and Students. **Third edition, revised and enlarged.** With 192 illustrations. 12mo, cloth...................$5.00

BOWIE, AUG. J., Jr., M.E. A Practical Treatise on Hydraulic Mining in California. With Descriptions of the Use and Construction of Ditches, Flumes, Wrought-iron Pipes and Dams; Flow of Water on Heavy Grades, and its Applicability, under High Pressure, to Mining. **Tenth edition.** Quarto, cloth. Illustrated...........................$5.00

BOWSER, E. A., Prof. An Elementary Treatise on Analytic Geometry. Embracing Plane Geometry, and an Introduction to Geometry of Three Dimensions. **Twenty-second edition,** 12mo, cloth..............$1.75

—— **An Elementary Treatise on the Differential and Integral Calculus.** With numerous examples. **Twenty-second edition,** enlarged by 640 additional examples. 12mo, cloth................................$2.25

—— **An Elementary Treatise on Analytic Mechanics.** With numerous examples. **Twentieth edition,** 12mo, cloth....................$3.00

—— **An Elementary Treatise on Hydro-Mechanics.** With numerous examples. **Sixth edition.** 12mo, cloth $2.50

—— **A Treatise on Roofs and Bridges.** With numerous Exercises. Especially adapted for school use. **Second edition.** 12mo, cloth. Illustrated. net, $3.00

BROWN, SIR HANBURY, K.C.M.G. Irrigation: Its Principles and Practice as a Branch of Engineering. 8vo, cloth, 301 pp. Illustrated. net, $5.00

BRUCE, E. M., Prof. Pure Food Tests: the Detection of the Com- mon Adulterants of Foods by simple Qualitative Tests. A ready manual for Physicians, Health Officers, Food Inspectors, Chemistry Teachers, and all especially interested in the Inspection of Food. 12mo, cloth, illustrated. net, $1.25

BRUHNS, Dr. New Manual of Logarithms to Seven Places of Decimals. **Seventh Edition.** 8vo, half morocco $2.50

CAIN, W., Prof. Brief Course in the Calculus. With figures and diagrams. 8vo, cloth, illustrated .... net, $1.75

CARPENTER, R. C., Prof., and DIEDERICHS, H., Prof. Internal Combustion Engines. With figures and diagrams. 8vo, cloth, illustrated .... net, $4.00

CATHCART, WM. L., Prof. Machine Design. Part I. Fastenings. 8vo, cloth. Illustrated .... net, $3.00

CHAMBERS' MATHEMATICAL TABLES, consisting of Logarithms of Numbers 1 to 108,000. Trigonometrical, Nautical and other Tables. **New edition.** 8vo, cloth $1.75

CHRISTIE, W. WALLACE. Chimney Design and Theory. A book for Engineers and Architects, with numerous half-tone illustrations and plates of famous chimneys. **Second edition, revised.** 8vo, cloth...$3.00

—— **Boiler Waters. Scale, Corrosion, Foaming.** 8vo, cloth. Illustrated .... net, $3.00

CORNWALL, H. B., Prof. Manual of Blow-pipe Analysis, Quali-tative and Quantitative. With a Complete System of Determinative Mineralogy. With many illustrations. 8vo, cloth $2.50

CROCKER, F. B., Prof. Electric Lighting. A Practical Exposition of the Art, for Use of Engineers, Students, and others interested in the Installation or Operation of Electrical Plants. **Eighth edition, thoroughly revised and rewritten. Vol. I. The Generating Plant.** 8vo, cloth. Illustrated $3.00

—— **Vol. II. Distributing Systems and Lamps. Sixth edition.** 8vo, cloth. Illustrated $3.00

—— **and WHEELER, S. S. The Management of Electrical Ma-**chinery. A thoroughly revised and enlarged edition of "The Practical Management of Dynamos and Motors." **Eighth edition, twenty-fourth thousand.** 131 illustrations. 12mo, cloth. 223 pp...net, $1.00

DINGER, H. C., Lieut., U.S.N. Handbook for the Care and Opera-tion of Naval Machinery. 16mo, cloth, illustrated .... net, $2.00

DORR, B. F. The Surveyor's Guide and Pocket Table-Book. Seventh edition, revised. and greatly extended. With a second appendix up to date. 16mo, morocco, flaps $2.00

DRAPER, C. H. Heat, and the Principles of Thermo-Dynamics. With many illustrations and numerical examples. 12mo, cloth $1.50

ECCLES, Dr. R. G. Food Preservatives; their Advantages and Proper Use. With an Introduction by E. W. Duckwall, M.S. 8vo, 202 pp., cloth. $1.00
    " Paper. .... .50

ELIOT, C. W., Prof., and STORER, F. H. A., Prof. Compendious Manual of Qualitative Chemical Analysis. Revised with the co-operation of the authors, by Prof. William R. Nichols. Illustrated. **Twenty-second edition.** newly revised by Prof. W. B. Lindsay and F. H. Storer. 12mo, cloth..$1.50

EVERETT, J. D. Elementary Text-Book of Physics. Illustrated. **Seventh edition.** 12mo, cloth.............................:..........$1.40

EWING, A. J., Prof. The Magnetic Induction in Iron and other metals. **Third edition, revised.** 159 illustrations. 8vo, cloth....$4.00

FANNING, J. T. A Practical Treatise on Hydraulic and Water- supply Engineering. Relating to the Hydrology, Hydro-dynamics and Practical Construction of Water-works in North America. 180 illustrations. **Seventeenth edition, revised, enlarged,** and new tables and illustrations added· 650 pp. 8vo, cloth...................................$5.00

FISH, J. C L. Lettering of Working Drawings. Thirteen plates, with descriptive text. Oblong, $9 \times 12\frac{1}{2}$, boards...................$1.00

FLEMING, J. A., Prof. The Alternate-current Transformer in Theory and Practice. **Vol. I. The Induction of Electric Currents.** 611 pp. **New edition,** illustrated. 8vo, cloth...................$5.00

—— **Vol. II. The Utilization of Induced Currents.** Illustrated. 8vo, cloth...$5.00

—— Electrical Laboratory Notes and Forms, Elementary and Advanced. 4to, cloth, illustrated.................................$5.00

—— A Handbook for the Electrical Laboratory and Testing Room. **Vol. I. Equipment, Resistance, Current, Electromotive Force and Power Measurement.** 538 pages. Illustrated. 8vo, cloth...net, $5.00

—— **Vol. II. Meter, Lamp, Cable, Dynamo, Motor and Transformer Testing.** 650 pages. Illustrated. 8vo, cloth...............net, $5.00

FOSTER, H. A. Electrical Engineers' Pocket-Book. With the Collaboration of Eminent Specialists. A handbook of useful data for Electricians and Electrical Engineers. With innumerable tables, diagrams, and figures. **Fifth edition, completely revised and enlarged.** Pocket size, flexible leather, elaborately illustrated with an extensive index and patent thumb index tabs. 1636 pp.........................$5.00

FOX, WM., and THOMAS, C. W., M.E. A Practical Course in Mechan- ical Drawing. With plates. **Third edition, revised.** 12mo, cloth.$1.25

GANT, L. W. Elements of Electric Traction, for Motormen and Others. Illustrated. 217 pp., 8vo, cloth....................net, $2.50

GEIKIE, J. Structural and Field Geology, for Students of Pure and Applied Science. With figures, diagrams, and half-tone plates. 8vo, cloth...net, $4.00

GILLMORE, Q. A., Gen. Practical Treatise on the Construction of Roads, Streets, and Pavements. **Tenth edition.** With 70 illustrations. 12mo, cloth..$2.00

GOODEVE, T. M. A Text-Book on the Steam-Engine. With a Supplement on Gas-Engines. **Twelfth edition, enlarged.** 143 illustrations. 12mo, cloth...$2.00

GUNTHER, C. O., Prof. Integration by Trigonometric and Imaginary Substitution. With an Introduction by J. Burkitt Webb. Illustrated. 12mo, cloth..net, $1.25

GUY, A. E. Experiments on the Flexure of Beams, resulting in the Discovery of New Laws of Failure by Buckling. Reprinted from the "American Machinist." With diagrams and folding plates. 8vo, cloth, illustrated. 122 pp...net, $1.25

HAEDER, HERMAN, C. E. A Handbook on the Steam-Engine. With especial reference to small and medium sized engines. **Third English edition,** re-edited by the author from the second German edition, and translated with considerable additions and alterations by H. H. P. Powles. Nearly 1100 illustrations. 12mo, cloth.................$3.00

HALE, W. J., Prof. (Univ. of Mich.) Calculations of General Chemistry, with Definitions, Explanations, and Problems. 174 pp. 12mo, cloth..net, $1.00

HALL, WM. S., Prof. Elements of the Differential and Integral Calculus. **Sixth edition, revised.** 8vo, cloth. Illustrated.....net, $2.25

———— Descriptive Geometry; With Numerous Problems and Practical Applications. Comprising an 8vo volume of text and a 4to Atlas of illustrated problems. **Second edition.** Two vols., cloth..........net, $3.50

HALSEY, F. A. Slide-Valve Gears: an Explanation of the Action and Construction of Plain and Cut-off Slide-Valves. Illustrated. **Eleventh edition, revised and enlarged.** 12mo, cloth.................$1.50

HANCOCK, HERBERT. Text-Book of Mechanics and Hydrostatics. With over 500 diagrams. 8vo, cloth.............................$1.75

HAWKESWORTH, J. Graphical Handbook for Reinforced Concrete Design. A series of plates, showing graphically, by means of plotted curves, the required design for slabs, beams, and columns under various conditions of external loading, together with practical examples showing the method of using each plate. 4to, cloth.................net, $2.50

HAY, A. Alternating Currents; Their Theory, Generation, and Transformation. 8vo, cloth, illustrated......................net, $2.50

———— Principles of Alternate-Current Working. 12mo, cloth, illustrated...$2.00

———— An Introductory Course of Continuous Current Engineering. With 183 figures and diagrams. Illustrated. 327 pp. 8vo, cloth.
net, $2.50

HECK, R. C. H., Prof. The Steam-Engine. Vol. I. The Thermodynamics and the Mechanics of the Engine. 8vo, cloth, 391 pp. Illustrated...net, $3.50

———— Vol. II. Form, Construction, and Working of the Engine. The Steam-Turbine. 8vo, cloth. Illustrated.....................net, $5.00

HERRMANN, GUSTAV. The Graphical Statics of Mechanism. A Guide for the Use of Machinists, Architects, and Engineers; and also a Text-Book for Technical Schools. Translated and annotated by A. P. Smith, M.E. 7 folding plates. **Sixth edition.** 12mo, cloth.....$2.00

HIROI, I. Statically-Indeterminate Stresses in Frames Commonly Used for Bridges. With figures, diagrams, and examples. 12mo, cloth, illustrated...net, $2.00

HOPKINS, N. MONROE, Prof. Experimental Electrochemistry. Theoretically and Experimentally Treated. 300 pp., 8vo. Illustrated. net, $3.00

HOUGHTON, C. E. The Elements of Mechanics of Materials. A text for students in engineering courses. Illustrated. 194 pp., 12mo, cloth...net, $2·00

HUTCHINSON, R. W., Jr. Long Distance Electric Power Trans- mission: being a treatise on the Hydro-electric Generation of Energy; its Transformation, Transmission, and Distribution. 12mo, cloth, illustrated, 345 pp...net, $3.00

JAMIESON, ANDREW, C. E. A Text-Book on Steam and Steam- Engines, including Turbines and Boilers. Specially arranged for the Use of Science and Art, City and Guilds of London Institute, and other Engineering students. **Fifteenth edition, revised.** Illustrated. 12mo, cloth. $3.00

——— **Elementary Manual on Steam, and the Steam-Engine.** Specially arranged for the Use of First-Year Science and Art, City and Guilds of London Institute, and other Elementary Engineering Students. **Twelfth edition.** 12mo, cloth...$1.50

JANNETTAZ, EDWARD. A Guide to the Determination of Rocks: being an Introduction to Lithology. Translated from the French by G. W. Plympton, Professor of Physical Science at Brooklyn Polytechnic Institute. **Second edition, revised.** 12mo, cloth................$1.50

JOHNSTON, J. F. W., Prof., and CAMERON, Sir CHARLES. Elements of Agricultural Chemistry and Geology. Seventeenth edition, 12mo, cloth...$2.60

KAPP, GISBERT, C. E. Electric Transmission of Energy, and its Transformation, Subdivision, and Distribution. A practical handbook. **Fourth edition, revised.** 12mo, cloth........................$3.50

KELLER, S. S., Prof. Mathematics for Engineering Students (Carnegie Schools Textbook Series. 12mo, half leather, illustrated.) Algebra and Trigonometry, with a chapter on Vectors. 282 pp...net, $1.75
——— Special Algebra Edition. 113 pp..........................net, $1.00
——— Plane and Solid Geometry, 212 pp..........................net, $1.25
——— Analytical Geometry and Calculus, 359 pp.................net, $2.00

KEMP, JAMES FURMAN, A.B., E.M. A Handbook of Rocks; for use without the microscope. With a glossary of the names of rocks and other lithological terms. **Fourth edition, revised.** 8vo, cloth. Illustrated...$1.50

KERSHAW, J. B. C. Electrometallurgy. Illustrated. 303 pp. 8vo, cloth..net, $2.00

KLEIN, J. F. Design of a High-Speed Steam-engine. With notes diagrams, formulas, and tables. **Second edition, revised and enlarged.** 8vo, cloth. Illustrated. 257 pp..................net, $5.00

KNIGHT, A. M., Lieut.-Com., U.S.N. Modern Seamanship. Illustrated with 136 full-page plates and diagrams. **Third edition, revised.** 8vo, cloth, illustrated.......................................net, $6.00
Half morrocco. $7.50

KOESTER, F. Steam-Electric Power Plants and their Construction. A Practical Treatise on the Design of Central Light and Power Stations and their Economical Construction and Operation. 473 pp., 340 illustrations. 8vo, cloth...net, $5.00

KRAUCH, C., Dr. Testing of Chemical Reagents for Purity. Authorized translation of the Third Edition, by J. A. Williamson and L. W. Dupre. With additions and emendations by the author. 8vo, cloth, net, $3.00

LAMBORN, L. L. Cottonseed Products. A Manual of the Treatment of Cottonseed for its Products and Their Utilization in the Arts. With tables, figures, full-page plates, and a large folding map. 8vo, cloth, illustrated..net, $3.00

LANCHESTER, F. W. Aerodynamics: Constituting the First Volume of a Complete Work on Aerial Flight. With Appendices on the Velocity and Momemtum of Sound Waves, on the Theory of Soaring Flight, etc. With numerous diagrams and half-tones. Illustrated. 442 pp , 8vo, cloth...net, $6.00

LASSAR-COHN, Dr. An Introduction to Modern Scientific Chem- istry, in the form of popular lectures suited to University Extension Students and general readers. Translated from the author's corrected proofs for the second German edition, by M. M. Pattison Muir, M.A. 12mo, cloth. Illustrated...$2.00

LATTA, M. N. Handbook of American Gas-Engineering Practice. With diagrams and tables. 8vo, cloth, illustrated, 460 pp..net, $4.50

LEEDS, C. C. Mechanical Drawing for Trade Schools. High School edition. (Carnegie Technical Schools Textbooks.) Text and Plates. 4to, oblong cloth...net, $1.25

———— **Mechanical Drawing for Trade Schools. Machinery Trades' edition.** 43 lessons in text and plates. 4to, oblong cloth,150 pp. net, $2.00

LIVERMORE, V. P., and WILLIAMS, J. How to Become a Com- petent Motorman. Being a Practical Treatise on the Proper Method of Operating a Street Railway Motor Car; also giving details how to overcome certain defects. **Revised edition, entirely rewritten and enlarged.** 16mo, cloth, illustrated.............................$1.00

LODGE, OLIVER J. Elementary Mechanics, including Hydrostatics and Pneumatics. **Revised edition.** 12mo, cloth $1.50

LUCKE, C. E. Gas Engine Design. With figures and diagrams. Second edition, revised. 8vo, cloth, illustrated.............net, $3.00

LUNGE, G., Ph.D. Technical Chemists' Handbook. Tables and methods of analysis for manufacturers of inorganic chemical products. 283 pp. 12mo, leather.....................................net, $3.50

LUQUER, LEA McILVAINE, Ph.D. Minerals in Rock Sections. The Practical Method of Identifying Minerals in Rock Sections with the Microscope. Especially arranged for Students in Technical and Scientific Schools. **Third edition, revised.** 8vo, cloth. Illustrated....net, $1.50

MASSIE, W. W., and UNDERHILL, C. R. Wireless Telegraphy and Telephony Popularly Explained. With a special article by Nikola Tesla. 76 pp. 28 illustrations. 12mo, cloth..................net, $1.00

MELICK, C. W., Prof. Dairy Laboratory Guide. 12mo, cloth, illustrated...net, $1.25

MERCK, E. Chemical Reagents: Their Purity and Tests. 250 pp...net, $1.50

MILLER, E. H. Quantitative Analysis for Mining Engineers. Second edition, revised. 8vo, cloth.............................net, $1.50

MINIFIE, WM. Mechanical Drawing. A Text-Book of Geometrical Drawing for the use of Mechanics and Schools, in which the Definitions and Rules of Geometry are familiarly explained; the Practical Problems are arranged from the most simple to the more complex, and in their description technicalities are avoided as much as possible. With illustrations for Drawing Plans, Sections, and Elevations of Railways and Machinery; an Introduction to Isometrical Drawing, and an Essay on Linear Perspective and Shadows. Illustrated with over 200 diagrams engraved on steel. **Tenth thousand.** With an appendix on the Theory and Application of Colors. 8vo, cloth..$4.00

MINIFIE, WM. Geometrical Drawing. Abridged from the Octavo Edition, for the use of schools. Illustrated with 48 steel plates. **Ninth edition.** 12mo, cloth.......................................$2.00

MOSES, ALFRED J., and PARSONS, C. L. Elements of Mineralogy, Crystallography, and Blow-Pipe Analysis, from a Practical Standpoint. 336 illustrations. **Fourth edition.** 8vo, cloth...............$2.50

NASMITH, JOSEPH. The Student's Cotton Spinning. Thirteenth thousand, revised and enlarged. 8vo, cloth. Illustrated.......$3.00

NUGENT, E. Treatise on Optics; or, Light and Sight theoretically and practically treated, with the application to Fine Art and Industrial Pursuits. With 103 illustrations. 12mo, cloth..................$1.50

OLSEN, Prof. J. C. Text-Book of Quantitative Chemical Analysis by Gravimetric, Electrolytic, Volumetric, and Gasometric Methods. With seventy-two Laboratory Exercises giving the analysis of Pure Salts, Alloys, Minerals, and Technical Products. **Fourth edition, revised and enlarged.** 8vo, cloth. Illustrated. 513 pp...................net, $4.00

OLSSON, A. Motor Control as Used in Connection with Turret Turning and Gun Elevating. (The Ward Leonard System.) Illustrated. 8vo, Pamphlet, 27 pp. (U. S. Navy Electrical Series, No. 1.)......net, $.50

OUDIN, MAURICE A. **Standard Polyphase Apparatus and Systems.** With many diagrams and figures. **Sixth edition, thoroughly revised.** Fully illustrated. 8vo, cloth $3.00

PALAZ, A., Sc.D. **A Treatise on Industrial Photometry,** with special application to Electric Lighting. Authorized translation from the French by George W. Patterson, Jr. **Second edition, revised.** 8vo, cloth. Illustrated..$4.00

PARSHALL, H. F., and HOBART, H. M. **Armature Windings of** Electric Machines. With 140 full-page plates, 65 tables, and 165 pages of descriptive letter-press. **Second edition.** 4to, cloth...........$7.50

—— **Electric Railway Engineering.** With numerous tables, figures, and folding plates. 4to, cloth, 463 pp., illustrated...........net, $10.00

PAULDING, CHAS, P. **Practical Laws and Data on Condensation** of Steam in Covered and Bare Pipes. 12mo, cloth. Illustrated. 102 pages..net, $2.00

—— **The Transmission of Heat through Cold-Storage Insulation.** Formulas, Principles, and Data relating to Insulation of every kind. A Manual for Refrigerating Engineers. 12mo, cloth. 41 pp. Illustrated. net, $1.00

PERRINE, F. A. C., A.M., D.Sc. **Conductors for Electrical Distribution**; Their Manufacture and Materials, the Calculation of the Circuits, Pole Line Construction, Underground Working and other Uses. With diagrams and engravings. **Second edition, revised.** 8vo, cloth...net, $3.50

PERRY, JOHN. **Applied Mechanics. A Treatise for the Use of** Students who have time to work experimental, numerical, and graphical exercises illustrating the subject. **New edition, revised and enlarged.** 650 pages. 8vo, cloth..................................net, $2.50

PLATTNER. **Manual of Qualitative and Quantitative Analysis with** the Blow-Pipe. From the last German edition, revised and enlarged. by Prof. Th. Richter, of the Royal Saxon Mining Academy. Translated by Prof. H. B. Cornwall, assisted by John H. Caswell. Illustrated with 78 woodcuts. **Eighth edition, revised.** 463 pages. 8vo, cloth. .net, $4.00

POPE, F. L. **Modern Practice of the Electric Telegraph.** A Technical Handbook for Electricians, Managers, and Operators. **Seventeenth edition, rewritten and enlarged,** and fully illustrated. 8vo, cloth.$1.50

PRELINI, CHARLES. **Tunneling.** A Practical Treatise containing 149 Working Drawings and Figures. With additions by Charles S. Hill, C.E., Associate Editor "Engineering News." **Third edition, revised.** 8vo, cloth. Illustrated...$3.00

—— **Earth and Rock Excavation.** A Manual for Engineers, Contractors, and Engineering Students. **Second edition, revised.** 8vo, cloth. Illustrated. 350 pp.....................................net, $3.00

—— **Graphical Determination of Earth Slopes. Retaining Walls,** and Dams. 8vo, cloth, illustrated, 136 pp..................net, $2.00

PRESCOTT, A. B., Prof. Organic Analysis. A Manual of the Descriptive and Analytical Chemistry of Certain Carbon Compounds in Common Use; a Guide in the Qualitative and Quantitative Analysis of Organic Materials in Commercial and Pharmaceutical Assays, in the Estimation of Impurities under Authorized Standards, and in Forensic Examinations for Poisons, with Directions for Elementary Organic Analysis. **Sixth edition.** 8vo, cloth...$5.00

———— and **SULLIVAN, E. C. First Book in Qualitative Chemistry. Eleventh edition.** 12mo, cloth...........................net, $1.50

———— and **OTIS COE JOHNSON.** Qualitative Chemical Analysis. A Guide in the Practical Study of Chemistry and in the Work of Analysis. **Sixth revised and enlarged edition, entirely rewritten.** with an Appendix by H. H. Willard containing a few improved methods of Analysis. 8vo, cloth..net, $3.50

RANKINE, W. J. MACQUORN, C.E., LL.D., F.R.S. Machinery and Mill-work. Comprising the Geometry, Motions, Work, Strength, Construction, and Objects of Machines, etc. Illustrated with nearly 300 woodcuts. **Seventh edition.** Thoroughly revised by W. J. Millar. 8vo, cloth...$5.00

—— —— **The Steam-Engine and Other Prime Movers.** With diagram of the Mechanical Properties of Steam. With folding plates, numerous tables and illustrations. **Fifteenth edition.** Thoroughly revised by W. J. Millar. 8vo, cloth...$5.00

———— **Useful Rules and Tables for Engineers and Others.** With appendix, tables, tests, and formulæ for the use of Electrical Engineers. Comprising Submarine Electrical Engineering, Electric Lighting, and Transmission of Power. By Andrew Jamieson, C.E., F.R.S.E. **Seventh edition, thoroughly revised** by W. J. Millar. 8vo, cloth........$4.00

———— **A Mechanical Text-Book.** By Prof. Macquorn Rankine and E. E. Bamber, C.E. With numerous illustrations. **Fifth edition.** 8vo, cloth...$3.50

RANKINE, W. J. MACQUORN, C.E., LL.D., F.R,S. Applied Mechanics. Comprising the Principles of Statics and Cinematics, and Theory of Structures, Mechanics, and Machines. With numerous diagrams. **Eighteenth edition.** Thoroughly revised by W. J. Millar. 8vo, cloth $5.00

———— **Civil Engineering.** Comprising Engineering, Surveys, Earthwork, Foundations, Masonry, Carpentry, Metal-Work, Roads, Railways, Canals, Rivers, Water-Works, Harbors, etc. With numerous tables and illustrations. **Twenty-third edition.** Thoroughly revised by W. J. Millar. 8vo, cloth...$6.50

RATEAU, A. Experimental Researches on the Flow of Steam Through Nozzles and Orifices, to which is added a note on The Flow of Hot Water. Authorized translation by H. Boyd Brydon. 12mo, cloth. Illustrated...net, $1.50

RAUTENSTRAUCH, W., Prof., and WILLIAMS, J. T. Machine Drafting and Empirical Design. A Textbook for Students in Engineering Schools and Others Who are Beginning the Study of Drawing as Applied to Machine Design. Part I. Machine Drafting. Illustrated, 70 pp., 8vo, cloth ....net, $1.25 Complete in Two Parts. Part II in preparation.

RAYMOND, E. B. **Alternating Current Engineering Practically** Treated. **Third edition, revised and enlarged,** with an additional chapter on "The Rotary Converter." 12mo, cloth. Illustrated. 232 pages. net, $2.50

REINHARDT, CHAS. W. **Lettering for Draughtsmen, Engineers and** Students. A Practical System of Free-hand Lettering for Working Drawings. New and revised edition. **Thirty-first** thousand. Oblong boards. $1.00

RICE, J. M., Prof., and JOHNSON, W. W., Prof. On a New Method of Obtaining the Differential of Functions, with especial reference to the Newtonian Conception of Rates of Velocities. 12mo, paper........$0.50

RIPPER, WILLIAM. **A Course of Instruction in Machine Drawing** and Design for Technical Schools and Engineer Students. With 52 plates and numerous explanatory engravings. **Second edition.** 4to, cloth.$6.00

ROBINSON, J. B. **Architectural Composition.** An attempt to order and phrase ideas which hitherto had been only felt by the instinctive taste of designers. 233 pp., 173 illustrations. 8vo, cloth...........net, $2.50

ROGERS, ALLEN. **A Laboratory Guide of Industrial Chemistry.** Illustrated. 170 pp. 8vo, cloth...........................net, $1.50

SCHMALL, C. N. **First Course in Analytic Geometry,** Plane and Solid, with Numerous Examples. Containing figures and diagrams. 12mo, half leather, illustrated.......................................net, $1.75

———— **and SHACK, S. M.** **Elements of Plane Geometry.** An Elementary Treatise. With many examples and diagrams. 12mo, half leather, illustrated..net, $1.25

SEATON, A. E., and ROUNTHWAITE, H. M. **A Pocket-book of** Marine Engineering Rules and Tables. For the Use of Marine Engineers and Naval Architects, Designers, Draughtsmen, Superintendents and all engaged in the design and construction of Marine Machinery, Naval and Mercantile. **Seventh edition, revised and enlarged.** Pocket size. Leather, with diagrams..................,.........................$3.00

SEIDELL, A. (Bureau of Chemistry, Wash., D. C.). **Solubilities of** Inorganic and Organic Substances. A handbook of the most reliable Quantitative Solubility Determinations. 8vo, cloth, 367 pp.....net, $3.00

SEVER, Prof. G. F. **Electrical Engineering Experiments and Tests** on Direct-Current Machinery. With diagrams and figures. **Second edition, thoroughly revised and enlarged.** 8vo, pamphlet. Illustrated..net, $1.00

———— **and TOWNSEND, F.** Laboratory and Factory Tests in Electrical Engineering. **Second Edition, thoroughly revised and enlarged.** 8vo, cloth. Illustrated. 236 pages.......................net, $2.50

SHELDON, S., Prof., and MASON, HOBART, B.S. **Dynamo Electric** Machinery; its Construction, Design, and Operation. Direct-Current Machines. **Seventh edition, revised.** 12mo, cloth. Illustrated. net, $2.50

SHELDON, S., MASON, H., and HAUSMANN, E. Alternating Current Machines. Being the second volume of the authors' "Dynamo Electric Machinery; its Construction, Design, and Operation." With many diagrams and figures. (Binding uniform with volume I.) **Seventh edition, completely rewritten.** 12mo, cloth. Illustrated..........net, $2.50

SHIELDS, J. E. Notes on Engineering Construction. Embracing Discussions of the Principles involved, and Descriptions of the Material employed in Tunneling, Bridging, Canal and Road Building, etc. 12mo, cloth..$1.50

SHUNK, W. F. The Field Engineer. A Handy Book of Practice the Survey, Location and Track-work of Railroads, containing a large collection of Rules and Tables, original and selected, applicable to both the standard and Narrow Gauge, and prepared with special reference to the wants of the young engineer. **Nineteenth edition, revised and enlarged.** With addenda. 12mo, morocco, tucks...........................$2.50

SMITH, F. E. Handbook of General Instruction for Mechanics. Rules and formulæ for practical men. 12mo, cloth, illustrated. 324 pp. net, $1.50

SOTHERN, J. W. The Marine Steam Turbine. A practical description of the Parsons Marine Turbine as now constructed, fitted and run, intended for the use of students, marine engineers, superintendent engineers draughtsmen, works managers, foremen, engineers and others. **Third edition, rewritten up to date and greatly enlarged.** 180 illustrations and folding plates, 352 pp. 8vo, cloth.......................net, $5.00

STAHL, A. W., and WOODS, A. T. Elementary Mechanism. A Text-Book for Students of Mechanical Engineering. **Sixteenth edition, enlarged.** 12mo, cloth..$2.00

STALEY, CADY, and PIERSON, GEO. S. The Separate System of Sewerage; its Theory and Construction. With maps, plates, and illustrations. **Third edition, revised and enlarged,** with a chapter on "Sewage Disposal." 8vo, cloth.................................$3.00

STODOLA, Dr. A. The Steam-Turbine. With an appendix on Gas Turbines and the future of Heat Engines. Authorized Translation from the Second Enlarged and Revised German edition by Dr. Louis C. Loewenstein. 8vo, cloth. Illustrated. 434 pages..................net, $4.50

SUDBOROUGH, J. J., and JAMES, T. C. Practical Organic Chem-istry. 92 illustrations. 394 pp., 12mo, cloth.................net, $2.00

SWOOPE. C. WALTON. Practical Lessons in Electricity. Principles, Experiments, and Arithmetical Problems. An Elementary Text-Book. With numerous tables, formulæ, and two large instruction plates. **Tenth edition.** 12mo, cloth. Illustrated...................net, $2.00

TITHERLEY, A. W., Prof. Laboratory Course of Organic Chemistry, including Qualitative Organic Analysis. With figures. 8vo, cloth. Illustrated..net, $2.00

THURSO, JOHN W. Modern Turbine Practice and Water-Power Plants. **Second edition, revised.** 8vo, 244 pages. Illustrated. net, $4.00

TOWNSEND, F. **Short Course in Alternating Current Testing.** 8vo, pamphlet. 32 pages..net, $0.75

URQUHART, J. W. **Dynamo Construction.** A practical handbook for the use of Engineer-Constructors and Electricians in charge, embracing Framework Building, Field Magnet and Armature Winding and Grouping, Compounding, etc., with examples of leading English, American, and Continental Dynamos and Motors. With numerous illustrations. 12mo, cloth..$3.00

VAN NOSTRAND'S Chemical Annual, based on Biederman's " Chiemker Kalender." Edited by Prof. J. C. Olsen, with the co-operation of Eminent Chemists. **Revised and enlarged.** Second issue 1909. 12mo, cloth..net, $2.50

VEGA, Von (Baron). **Logarithmic Tables of Numbers and Trigonometrical Functions.** Translated from the 40th, or Dr. Bremiker's thoroughly revised and enlarged edition, by W. L. F. Fischer, M.A., F.R.S. **Eighty-first edition.** 8vo, half morocco......................$2.50

WEISBACH, JULIUS. **A Manual of Theoretical Mechanics.** **Ninth American edition.** Translated from the fourth augmented and improved German edition, with an Introduction to the Calculus by Eckley B. Coxe, A.M., Mining Engineer. 1100 pages, and 902 woodcut illustrations. 8vo, cloth..$6.00
Sheep..$7.50

―――― **and HERRMANN, G.** **Mechanics of Air Machinery.** Authorized translation with an appendix on American practice by Prof. A. Trowbridge. 8vo, cloth, 206 pages. Illustrated..............net, $3.75

WESTON, EDMUND B. **Tables Showing Loss of Head Due to** Friction of Water in Pipes. **Fourth edition.** 12mo, full leather...$1.50

WILLSON, F. N. **Theoretical and Practical Graphics.** An Educational Course on the Theory and Practical Applications of Descriptive Geometry and Mechanical Drawing. Prepared for students in General Science, Engraving, or Architecture. **Third edition, revised.** 4to, cloth, illustrated..net, $4.00

―――― **Descriptive Geometry, Pure and Applied, with a chapter on** Higher Plane Curves, and the Helix. 4to, cloth, illustrated.....net, $3.00

WILSON, GEO. **Inorganic Chemistry, with New Notation.** Revised and enlarged by H. G. Madan. **New edition.** 12mo, cloth.....$2.00

WINCHELL, N. H., and A. N. **Elements of Optical Mineralogy.** An introduction to microscopic petrography, with descriptions of all minerals whose optical elements are known and tables arranged for their determination microscopically. 354 illustrations. 525 pages. 8vo, cloth..net, $3.50

WRIGHT, T. W., Prof. **Elements of Mechanics,** including Kinematics, Kinetics, and Statics. **Seventh edition, revised.** 8vo, cloth.....$2.50

―――― **and HAYFORD, J. F.** **Adjustment of Observations by the** Method of Least Squares, with applications to Geodetic Work. **Second edition, rewritten.** 8vo, cloth, illustrated..................net, $3.00

ZEUNER, A., Dr. **Technical Thermodynamics.** Translated from the Fifth, completely revised German edition of Dr. Zenner's original treatise on Thermodynamics, by Prof. J. F. Klein, Lehigh University. 8vo, cloth, two volumes, illustrated, 900 pages ....net, $8.00

RETURN TO ➡ CIRCULATION DEPARTMENT
202 Main Library

LOAN PERIOD 1 **HOME USE**	2	3
4	5	6

ALL BOOKS MAY BE RECALLED AFTER 7 DAYS
1-month loans may be renewed by calling 642-3405
6-month loans may be recharged by bringing books to Circulation Desk
Renewals and recharges may be made 4 days prior to due date

DUE AS STAMPED BELOW

AUG 08 1992		

UNIVERSITY OF CALIFORNIA, BERKELEY
FORM NO. DD6, 60m, 12/80   BERKELEY, CA 94720

CPSIA information can be obtained
at www.ICGtesting.com
Printed in the USA
BVHW08*1526041018
529297BV00008B/169/P